THE START OF AN EARTHMOV

The history of JCB. goes back over 45 years to 1945, when Joseph Cyril Bamford left the family firm of 'Bamfords Ltd' in Uttoxeter, Staffs. and on the day his first son Anthony was born, set up in business on his own as an Agricultural Engineer. He bought a second hand welding set and rented a small 12' x 18' garage for 30/- a week and he was ready for business. The first JCB. product was an old farm trailer which Joe Bamford rebuilt, painted, and offered for sale at Uttoxeter Market for £90. Unfortunately no one was interested, and later it was sold for £45. and an old trailer was taken in part exchange. Joe set about renovating this and later this was sold for £45.

In 1946 the first JCB four wheeled trailer was built, and Joe also started dealing in reconditioned second hand ex-Ministry vehicles which were in very great demand. JCB by now had out-grown the small garage and moved to an old coach house block at 'Crakemarsh Hall' between Uttoxeter and Rocester. Here they started to build 'Screw type' tipping trailers with materials salvaged from dismantled air-raid shelters. Joe Bamford was now employing Arthur Harrison and Bill Hirst along with several part-time workers.

By 1948 he was building hydraulic tipping trailers out of materials salvaged from 'Anderson' air-raid shelters, and also started refurbishing war time "Jeeps" adding wooden bodies (supplied by a local undertaker!) to turn them into shooting brakes. These proved to be a very profitable line, as vans and cars were still in short supply after the war.

JCB 'MAJOR' Loader:

After recognising the possibilities of hydraulic rams for lifting and lowering Joe designed and produced the 'Major Loader' in 1949 for fitting to 'Fordson E27N' and 'Nuffield Universal' tractors. The Major Loader was the first industrial hydraulic front end loader in Europe. It was a well built loader, and was suitable for the many new 'Industrial and Construction' markets, as well as for Forestry and Agricultural use. It was fixed with 2 single acting hydraulic lift rams, and had a maximum lift height of 12' 9" (with bucket fitted).

The JCB 'Major Loader' was built mainly of four two inch channel (again salvaged from ex air-raid shelters), and could lift to its maximum height in 7 seconds. A 25 gallon hydraulic oil tank was fitted to the right hand side and replaced the Fordson internal hydraulic pump. A 5/8 cubic yard bucket, muck fork bulldozer blade and crane attachment were available for the Major Loader. The 'Major Loader' remained in production until March 1957, when it was replaced by the JCB 'Loadall.

PRICES

Prices for the Major Loader in 1950 were as follows:-

Major Loader (fitting for Nuffield or Fordson)	£157
High Speed Pump	£115
General purpose bucket	£ 31
Bucket teeth	£ 5
Dozer blade	£ 65
15 cwt. Crane jib	£ 17
Ballast block	£ 17

Fitment by dealer
J C B cab complete

The trailer and Major Loader were all finished in dark green until 1950, when it changed to red. By 1954 it had changed to 'Cardinal Red' and from 1957 it changed again to bright yellow and has since remained unchanged but the shade of yellow is now darker.

Joe Bamford was working flat out, 7 days a week to produce the trailers and loaders, as demand was outstripping manufacturing capacity, but his Landlady at 'Crakemarsh Hall' disapproved of Sunday working so he moved not far away to the small village of Rocester, into a former cheese factory and pig farm. The building and outhouses were converted into an ideal workshop and outside storage yard, which has, over the years. expanded to become the present day factory site.

JCB MASTER LOADER.

In 1951 Joe Bamford's second son Mark was born, and in the same year the JCB Master Loader was introduced. The Master Loader was a smaller version of the Major Loader (which continued in production). The Loader sold for £68 mainly to Agricultural markets. It also had the same single acting rams of the Major Loader. The JCB Master Loader was not as successful as the Major Loader, and was only available to fit the Fordson Major E27N, but nevertheless sold in good numbers.

Joe Bamford also designed and started to manufacture a mid mounted hay mower for fitting to the Fordson and Ferguson tractors but this idea was sold off to Bamfords Ltd. of Uttoxeter and they produced and sold them for many years, marketing them under their own name. Joe was not in the position to commence the manufacture of the hay mowers, because he was already finding it difficult to complete orders for the Major Loader and trailers.

JCB HALFTRACK CONVERSIONS.

In 1952 when halftrack conversions of Agricultural tractors were becoming popular Joe Bamford designed and built a halftrack conversion for Fordson, Nuffield, David Brown, Allis Chalmers and Ferguson tractors. The conversion, based on the American 'Bombardier' half-track sold for £185. Also available to complete the conversion, was a set of skis to replace the front wheels. The original rear wheels and tyres were retained, and an idler axle and wheels were added in between the front and rear wheels, a rubber belt with steel plates acted as the track.

Production of halftrack conversions was discontinued as they were not sold in the large numbers to British farmers as JCB had expected. However, the conversions proved a very popular export to Scandinavian countries (three Ferguson tractors were fitted with JCB halftracks complete with skis and were driven to the South Pole in January 1958 by Sir Edmund Hilary. (One of the tractors is in the 'Massey-Ferguson' Tractor Museum at the Massey Ferguson Banner Lane plant in Coventry). Production was discontinued during 1957/58.

JCB SI-DRAULIC LOADER.

Joe Bamford introduced the 'Si-draulic' Loader in 1953. The Loader consisted of one left hand side mounted loader boom, with a front mounted fork or bucket, positioned centrally at the front of the tractor. It sold for £75. The rather novel idea did not appeal to British Farmers as it was unstable at full lift height, however, the Si-draulic Loaders were very popular in France, where 2,000 were built and sold under licence by Hydro Fourche. The Loader was not as popular in the U K and only a few were made by JCB when production ceased in 1956.

JCB TRADE MARK/LOGO.

The World famous JCB logo first appeared in 1953 on the JCB Loadover (JCB are the initials of Joseph Cyril Bamford). The logo was based around the side swath bar of the 1951 JCB Hay mower, which due to the shape of the bar, the letters were all of different heights. This was turned around to become JCB logo.

JCB LOADOVER.

The Loadover was a heavy industrial type loader built onto a Fordson 'New' Major tractor, and mainly sold to Gas and Coal Boards in the U K via. Knutsford motors, it is believed that less than a dozen were built. The 'Loadover' was another of Joe's innovative ideas. The Loader shovel could be filled and lifted up and right over the tractor to tip at the rear of the tractor, thus cutting down the need to shunt around reversing and turning to load a lorry. The first Loadover was built using parts from the J C B Major Loader and fitted onto a Fordson E27N tractor. The shovel was mounted over the rear wheels of the tractor and when lifted would swing over the top of the tractor and tip the shovel behind it.

The Loadover, mounted on a Fordson 'New' Major tractor in 1954 cost £1,494.

JCB'S FIRST EXCAVATOR.

In early 1953 whilst Joe Bamford was on a sales trip to Scandinavia (selling his JCB Halftrack conversions), he saw a 'Broyt' trailed Excavator working in Norway, which took his eye. Although crude in its design, build and operation, Joe could see the possibilities of improving both its design and operation. If he were to add the hydraulic rams that were used in his Major Loader and by adding a 180 degrees slew (instead of the 90 degrees slew of the Broyt).

Joe Bamford could forsee that a hydraulic excavator, would be a certain winner, and promptly purchased a Broyt excavator, which he had crated up and shipped back to England.

On his return, Joe uncrated the excavator and spent many hours using the machine, cleaning ditches, excavating trenches and generally finding out what it could do, and what its faults and failings were. Then it was off the drawing board to put his ideas down on paper.

The JCB Excavator Was Based on an 'A' Frame Attached to the tractor via. the 3 point linkage (being raised and lowered by the tractor hydraulic lift). The hydraulic pump fitted directly onto the tractor PTO shaft.

A flip up seat was attached near the top of the 'A' frame and the levers and valve block were located at the top of the 'A' frame with a small hydraulic tank and two filters fitted in the middle of the 'A' frame.

The boom was made of tubular steel, and the dipper manufactured from square section steel, was operated via a linkage bar and hydraulic ram, which in design was similar to the USA built Shawnee digger; later made under licence by Steelfab as the Shawnee 'Scout'.

The scoops (buckets) were available in 10", 18" and 24" widths, and could be fitted either facing the tractor or as a face shovel (facing away from the tractor. This unit was sold as the first JCB hydraulic excavator/trencher.

Maximum digging depth was 9'3" with 13' 7" reach and the machine sold for £850 ex works for fitting to a Fordson Major, Nuffield, or TE-20 tractor.

Due to the light weight of the excavator no front ballast weight was needed but only a handful of this, JCB's first hydraulic trencher were built before the updated production model JCB Mark 1 excavator was produced.

The Mk. 1 excavator was not the first excavator Joe Bamford produced as later histories would have us believe. The only known example of a JCB hydraulic trencher to survive is kept at the JCB factory awaiting restoration by JCB apprentices. The light weight of the first hydraulic excavator/trencher soon prompted its replacement in 1954 with a stronger and larger version.

THE JCB MKI. EXCAVATOR.

The result was the JCB Mk.I Hydraulic Excavator', which he had fitted onto the rear a Fordson 'New' Major tractor and Major Loader, (thereby creating Europe's first purpose built Excavator/Loader).

The Mk.I excavator had a maximum digging depth of 11ft. with a maximum reach of 15ft. A choice of 4 excavator buckets (originally called 'scoops'). was available in sizes of 18", 24", 30" and 48" and it was painted 'Cardinal' red.

The JCB Mk.I Hydraulic Excavator was completely hydraulically operated (unlike the Broyt which had been operated using wires and pulleys).

The tractor engine drove a front mounted 'Plessey' hydraulic pump which took hydraulic oil from a secondary external tank, mounted on the right hand side of the tractor. This oil was fed into the valve block and bank of 5 levers at the rear of the tractor on the excavator 'A' frame, where the driver could - via the 5 levers - activate the double acting hydraulic rams used to lift and lower the 'A' frame (to jack the rear of the tractor into its working position), to slew the back actor arm through a 180 degrees arc, lift and lower the boom and dipper, and to open and close the excavator bucket. Once Joe Bamford had perfected the design, the J C B Mk.I Excavator was put into production, in early 1954, starting at Serial No. 2001.

9 Mk.I Excavators were built and sold during the first month of production.

The JCB Mk.I Excavator was first shown and demonstrated locally in Staffs and West Midlands, and later at Agricultural shows, and in adverts in the Farming Press, which brought in enough orders to keep the small workforce busy for the rest of the year.

Price for the J C B Mk.I Excavator fitted to a Fordson Major or Nuffield tractor was £1,833 ex. works. Approximately 550 were built between 1954 and March 1957, when production ceased. At this time the workforce consisted of Joe and 5 part time workers with Joe's wife Meg seeing to the bookwork and administration.

THE FIRST JCB EXCAVATOR/ LOADER.

Since Joe was already building the Major Loader and fitting it to Nuffield and Fordson tractors, it was a natural progression for him to fit the Mk.I excavator on to the rear of the tractor to produce an Excavator/Loader unit, (the first in Europe). This useful combination soon became popular, and within a year a metal cab was available for an extra £52. During the production run from 1954 to March 1957 approximately 550 Mk.I excavators were built, many being exported, together with over 1,000 Major Loaders.

JCB "HYDRA-DIGGA"

Following the success of the Mk.I Excavator, Joe soon discovered that the market also required a larger version, and in 1956, (2 year after the Mk.I was first built) the 'JCB Hydra-digga' was produced. The 'A' frame mounting was similar to the Mk.I, but a larger boom and dipper was fitted, and a different mounting arrangement for the dipper ram was designed together with larger capacity excavator buckets. The Hydradigger was not a replacement for the Mk.I; it was sold as a larger version of the JCB Mk.I Excavator. The 'Hydra-digga' was available as a single unit, or as an Excavator/Loader with the Major Loader fitted with a choice of Fordson or BMC Nuffield tractor unit. Maximum digging depth was 13' and maximum reach 18'; as had been the case with the JCB Mk. 1 excavator, a steel cab was available

HYDRA-DIGGA AND MAJOR LOADER:
(Specifications and price 1957).

The Fordson diesel tractor, with 750 x 16 front and 14 x 30 rear wheel and tyres, drivers cab, electric lights, starter and hand brake, 5/8 Cu. yd. loader, bucket, bulldozer blade, and conversion operating valves, rear mounted fully hydraulic 'J C B Hydradigger' Excavator, fitted at works, tested and with oil £1,890 ex. works.

HYDRA-DIGGA AND BALLAST WEIGHT

Fordson Major diesel tractor with 750 x 16 front and 14 x 30 rear wheels and tyres, drivers cab, electric lights, starter and hand brake, rear mounted fully hydraulic JCB 'Hydra-digga' Excavator fitted at works and tested £1,164 ex. works.

Both the above machines did not come with Excavator buckets. These were available in different sizes, prices of buckets were as follows:-

18" Bucket	£48
24" Bucket	£52
30" Bucket	£63
46" Bucket	£89
12" Bucket (with ejector plate)	£65
Ditching Bucket	£55
Operators cab	£52
Ballast weight	£20

When the JCB Hydra-digga was introduced in March 1957 it was painted yellow with red buckets, (as opposed to the red colour scheme of the Mk.I Excavator). When factory fitted to Nuffield/Fordson tractors, the complete machine was painted yellow, but with red wheels and buckets.

Although when introduced in 1957, the JCB Hydra-digga was originally only available with fittings for Nuffield or Fordson tractors, by 1958/59 mounting frames were available for many popular crawlers, such as Marshall, IH Drott, County Crawler and County Fourdrive.

JCB HYDRA-DIGGA/IH B450

In April 1958 JCB produced a version of the Hydra-digga fitted onto the International Harvester 'B 450' tractor (with 55 horse power diesel engine).

The JCB Hydradigger/IH B450 could be supplied with cab, ballast block or Loadall 75 Front Shovel.

The Excavator was marketed solely by the International Harvester Company of London for sale via. IH Dealers in Britain and Europe, very few are believed to have been sold.

MAJOR LOADER PRICES 1956.

Major Loader fitted with 5/8 Cu. yd. bucket including high speed pump and Fordson Major tractor	£946
Ballast block (when excavator not fitted)	£35
Bulldozer blade	£68
High speed pump (loader type only)	£121
Conversion operating valve (enables loader to work off the excavator high speed pump)	£23
Fordson hydraulics and PTO.	£65
Bucket teeth (3)	£3
Crane jib	£18

LOOSE MACHINE FITTINGS

JCB Hydra-digga	£900
JCB Major Loader	£165
Loader Bucket 5/8 Cu. yd.	£33
Bonnet hinge for Fordson tractor	£5
Cab (Fordson or Nuffield)	£52

FITTING CHARGES AND PRICES AUGUST 1957

JCB Hydra-digga to Fordson tractor	£21
JCB Hydra-digga to Nuffield tractor	£31
JCB Hydra-digga to County four drive	£75
JCB Hydra-digga to County Crawler	£75
JCB Hydra-digga to I H BTD6 Crawler	£75
JCB Major Loader to Fordson/Nuffield tractor	£5

The first JCB Companies were registered in 1956. They were:-
1. J C Bamford (Excavators) Ltd. Directors: Joseph and Margaret Bamford.
2. Rocester Services Ltd., Directors: Joseph Bamford and Florence Margaret Wood.

J C Bamford (Excavators) Ltd. was responsible for the manufacturing and sales, and Rocester Services Ltd. saw to all the warranty claims and replacement parts.

LOADALL POWER SHOVEL

During 1955 Joe Bamford's 'Excavator Empire' was expanding quickly thanks to the success of the JCB Mk.I Excavator which was giving excellent service to both builders and farmers, especially when fitted with the Major Loader, to provide the ideal machine for digging, loading, lifting and levelling. With his order books full, Joe set about designing a replacement for the Major Loader.

Although the Major Loader was still popular, and selling well, Joe was unhappy about some parts of its design and to keep it competitive and ahead of competition, it needed strengthening, he also disliked the externally mounted hydraulic tank.

In 1958 the JCB 'Loadall' power shovel was introduced, with its innovative hydrachassis. The Loaders hollow mounting frame was enlarged to become the oil tank, providing stability, strength and a neat appearance, the Loadall's front bucket was hydraulically tipped and crowded via. a double acting ram. (unlike the JCB Major Loader which had a wire trip mechanism), the Loadall was available with fittings for Nuffield, Fordson Major with a smaller version for the Ferguson TE20 tractor. In 1959 the Loadall was re-named the Loadall 65, and a larger version 'The Loadall 75' was introduced. (The new Loadall 75, when fitted with the JCB Hydra-digga, founded the basis for the JCB 4 Excavator/Loader introduced in 1960).

A few of the last years production of JCB Hydra-digga (1959) had a new 2 lever valve block for all digging operation. Previously there were 4 levers for operating individual rams. and 1 for lifting and lowering the 'A' frame. It was not until 1961 that the 2 lever valve block was fully perfected and introduced.

The JCB 4 & 4C

The success of the JCB Hydra-digga/Loadall had during 1959 prompted the development for a larger more powerful machine.

The JCB 4 was introduced in 1960 to replace the Hydradigger and Loadall. The new machine incorporated an updated Loadall type front shovel and a strengthened version of the Hydra-digga, but it included built in mudguards, carrying hydraulic oil reservoir and a large 'JCB Superview' cab. Another new feature of the JCB 4 was a specially built front axle and yolk, built around a Fordson Power Major, (until 1960 when the Super Major skid unit was used). Maximum reach of the excavator was 19ft combined with a maximum digging depth of 13ft.

The JCB 4 had larger capacity excavator buckets than the Hydra-digga, due mainly to the larger bucket ram being fitted, and new 2 lever digging controls, (introduced in 1959 on the Hydra-digga).

The JCB 4 proved very popular with public works contractors, many could be seen at work on motorway projects in England during the early 1960's.

In 1962 the JCB 4 was updated by fitting longer reach front shovel arms, square mudguards and a new style cab to become the JCB 4C. Price in 1962 was £2,840.

JCB 3

The JCB 4 had been an immediate success, but had left a gap in the market for a smaller machine like the old JCB Excavator/Major Loader, which had ceased production in 1957.

The JCB 3 Excavator/Loader was first shown at the Crystal Palace Plant Show in the Summer of 1961, but did not go into production until November 1961. The new Excavator/Loader was built around a Fordson Super major tractor of 57 bhp. Although powered by the same tractor skid as the JCB 4 the 3 was much smaller and lighter, and subsequently more manoeuvrable than the 4.

The novel part about the JCB 3 was the JCB 'Hydrachassis' which consisted of a welded steel box section frame containing fuel tank, hydraulic oil tank, front axle yolk, front shovel mounting frame and a rear mounting bracket and stabiliser legs. The frame at the rear consisted of a series of cross rails to which the king post could be positioned along at any point. This became known as JCB 'Hydraslide' and 'Slydeze' king post. (All the other JCB Excavators had a centrally mounted king post and could not be moved either side).

Maximum digging depth of the JCB 3 was 11ft with 15ft maximum reach, The kingpost could be offset 2'10" from the central position.

The JCB Super Comfort cab complete with lights and indicators was available as an extra. By mid 1962 the JCB 3 round mudguards were changed for the square type, similar to the JCB 4. Price of the JCB 3 in 1962 was £2,500 (including optional cab).

JCB 1

The small JCB 1 was put on the market in 1962. Powered by a Petter PH2 cylinder diesel engine of 20 HP with 3 forward and 1 reverse gears. The machine consisted of a small version of the Hydradigger (but with 2 stabilisers similar to the JCB 3). A bulldozer blade was fitted to the front instead of a front shovel. The JCB 1 could dig 7ft. 6" and had a 10ft reach. It was 6ft wide and 18ft long. About 700 were made. It was aimed at the small builder, but was not a success. However, many were sold to Local Authorities for grave digging. A cab was available from 1963.

JCB 3C

By 1963 JCB had sold 7,000 excavators and were recognised as market leaders, and to press home the point, JCB introduced their most famous model, the JCB 3C.

It was a larger almost identical version of the JCB 3 but with longer reach front shovel arms, larger capacity front shovel, longer dipper arm and more spacious cab. Powered by the same Fordson Super Major tractor skid as the JCB 3 and 4C. It had a maximum reach of 18ft and maximum digging depth of 13ft. The 3C front loader maximum lift height was 13ft. 4".

Cabs for the JCB tractor shovels and Excavator/Loader from 1955 to 1968 were not manufactured by JCB, they were brought in from outside manufacturer 'S. Whiteley & Son' of Cleckheaton, Yorkshire and the wheels were made by Sankey of Wellington, Shropshire.

JCB 2 & 2B

The JCB 2 was smaller than the JCB 3 but still used the Fordson Super Major skid unit. It could dig to 10ft with maximum reach of 16ft. The JCB 2 weighed 2½ ton. It was later joined by the JCB 2B, which had a detachable rear hydraslide and back actor, the option of the tractor 3 point linkage could be fitted to allow the machine to be used for normal farm work, primarily aimed for local council and farm contractors use. The JCB 2 and 2B were in production from 1964 to 1968. Although sales were rather disappointing. A loader only version, the JCB 2L was also produced, and had a counter weight and towbar at the rear.

JCB DUMPERS.

November 1964 JCB introduced the 1D range of two wheel drive dumpers. The three dumpers came in 12 cwt., 22½cwt and 30 cwt. versions. The new dumpers were produced for the construction and house building industry to strengthen JCB's lead in that market, and to diversify into other markets. The dumpers were finished in the same colour as the Excavators, yellow with red wheels.

Specifications and prices for 16th. November 1964 as follows

30 CWT. DUMPER:

30 cwt. capacity, powered by JCB air cooled diesel engine, 20 bhp maximum at 2,000 rpm. 750 x 20 x 8 ply front, 600 x 16 x 4 ply rear tyres, 3 speed gear box and disc brakes. £610

22½ CWT. DUMPER:

22½ cwt. capacity dumper, powered by JCB air cooled diesel engine of 10 bhp maximum at 2,000 rpm. £469

15 CWT. DUMPER:

15 cwt. capacity dumper powered by JCB diesel engine of 10 bhp maximum at 2,000 rpm. £380

Production of the dumpers was not a success and they were phased out after 2 years.

DEMISE OF THE FORDSON SKID UNIT

In January 1965 JCB started having problems obtaining Fordson skid units (for the Excavator/Loader) from the Ford Motor Co. This was because Fords were introducing their range of Ford 'Force' tractors to replace the ageing Fordson EIA tractor. Joe Bamford flew to America with Mr. G G Buckley, General Manager of the new Ford tractor factory at Basildon, Essex to see R J Hampson, Vice Chairman of Ford Motor Co. (USA) to try to get continued supplies of Fordson skid units. Faced with the possible shut down of the JCB factory, Joe decided to change over to BMC Nuffield skid units. Joe Bamford held a celebration near the factory for Dealers, Press and selected customers, and to show his annoyance publicly by ceremoniously 'burying' a Fordson skid unit, to the delight of the invited guests, and to the annoyance of the Ford Motor Co!

Most of the guests thought this was hilarious, however, the Ford Motor Co. were not so amused! and there is no record of their views on the 'funeral'!

BMC SKID UNITS

The replacement for the Fordson Major skid units was BMC Nuffield 10/60 and 10/42 tractors developing 60 HP and 42 HP from 4 and 3 cylinder diesel engines built at the BMC factory at Bathgate in Scotland. The announcement of the change over was heralded as a vast improvement over the Fordson skid, for both power and ease of servicing and maintenance.

JCB 3D & 4D EXCAVATOR LOADERS 1967

The JCB 3D and 4D Excavator/Loaders were introduced in 1967. They were both powered by BMC Nuffield 10/60 skid units, of 60bhp at 2,200 rpm from a diesel engine. At 60HP the skid unit chosen by JCB was underpowered for the work that was required of it; nothing else suitable was available at the time. The design of the JCB 3D and 4D was based on the JCB 3 and JCB 3C Excavator/Loaders, but were neater looking machines, the 3D Excavator/Loader had a maximum digging depth of 15ft 1", and had a maximum reach of 19ft 11". The total weight of the 3D was 7 tonnes (less the weight of Excavator bucket). The JCB 4D replaced the 4C. (From 1968 the BMC 4/65 skid unit was used). The JCB 4D had a maximum digging depth of 16ft and a maximum reach of 20ft 9". It weighed 7 tonnes 12 cwt.

The old Mk.I JCB 2B, 3 and 3C were all updated and re-styled in the same style as the new 3D and 4D. They were known as the JCB MkII range.

The JCB 4D was too large and cumbersome, and production had ceased by 1971.

JCB MKII 180 DEGREE EXCAVATORS 1968-76

The JCB Mk. II range of Excavator/Loaders were introduced in 1968. The new machines were all similar in size to the J C B 2, 2B 3 and 3C Excavator/Loaders which they replaced. They even retained the model numbers, but the styling of the MkII diggers were based on the 3D and 4D introduced the previous year. The new models were the JCB 2B, 3 and 3C Mk. II. The 2B, 3, 3C, 3D and 4D Excavator/Loaders had many inter-changeable parts, which were common to all models in the Mk. II range and which kept dealers parts stock to a minimum. The MkII diggers were still painted in JCB yellow, with yellow cab, whilst the wheels and buckets were painted red.

The JCB 3C, 3D and 4D were built around the BMC 4/65 skid unit it was powered by a BMC 4 cylinder diesel engine of 65 bhp @ 2,200 rpm. which was basically an updated version of the BMC 10/60 skid unit.

The 2B and 3 were built around the BMC 3/45 skid unit, the engine was a BMC 3 cylinder diesel engine of 45 bhp. It was a similar type engine to the one fitted to the 3D and 4D and 3C, and shared many common parts with the 4 cylinder engine.

The JCB 2 (which was identical to the 2B, but did not have the detachable backhoe) was dropped from the new range, as JCB felt that future sales would be too small to continue production based on results over the previous few years.

The new JCB 180 degree Excavator/Loader range were well received by builders, contractors, and plant hire companies when they were launched in 1968. Prices for the new MK. II range were as follows:

J C B 2B   £3,152 (£3,260 with 3 point linkage)
J C B 3   £3,340
J C B 3C   £3,852
J C B 3D   £4,060
J C D 4D   £4,195

A torque convertor transmission was also available for the MK. II Excavator/Loaders. The Leyland 4/98 diesel engine of 70 bhp was fitted to the J C B 2D, 3, 3C and 3D 180 degree Excavator/Loaders in December 1970. It replaced the 65 bhp BMC engine. when the JCB 2B was fitted with the 70 bhp engine it was renamed the 2D. The JCB Excavator/Loaders received a number of improvements during the first few months of 1973, including a cerametallic clutch which replaced the sintered bronze clutch. A white cab on all machines, and a sliding side window. An improved heavy duty front axle, and the instrument panel was changed.

By 1975 JCB had over 50% of the U K 180 degree Excavator/Loader market, and was exporting machines to over one hundred countries world wide, (J C B is now a market leader in over 55 countries).

JCB MKIII RANGE.

The JCB Mk. II range of Excavator/Loaders were replaced in February 1977 by the Mk. III range. The J C B 2D, 3, 3C and 3D were similar in appearance to the Mk.IIs as most of the differences were not immediately visible. The cab was 6" higher than the Mk. II and had an opening lower half below the sliding door to allow easier access. The front bucket had a rounded profile inside.

In 1978 the JCB 'Powertrain', also known as 'build 5' 3CIII Excavator/Loaders were introduced, they had a Leyland 72 bhp diesel engine and the same torque convertor/transmission rear axle as the 3CX machines, this was partly to try the transmission out and to get people used to it, ready for the launching of the 3CX. Cab trim on the 'Powertrain' JCB models were also of a higher standard.

On the JCB Mk. III range of Excavators/Loaders, the optional extras such as 6 in 1 Clam shovel and extending dipper, were being publicised more by JCB Salesmen then had been the case in the past. (This was due mainly to influence from competitors such as Case and Massey Ferguson), who were starting to affect JCB's U K sales lead. In many European Countries during the Late 1970's JCB's were thought of as being too heavy and lacking in cab and driver comfort. The JCB 3C III was still their best seller, but it was becoming obvious that a replacement would soon be needed, in October 1980 the long awaited replacement was first launched, in the form of the JCB 3CX and JCB 3D (x). It was at this time that JCB dropped the JCB 2D and 3 Excavator/Loaders.

J C B 3CX. 3D 180 degree EXCAVATOR/LOADERS.

In October 1980 JCB announced their new range of 180 degree Excavator/Loaders. They were the JCB 3C, 3D (and 3DS Loading Shovel) all the new models were based on a Leyland 272 skid unit powered by a 4/98, Leyland 4 cylinder diesel engine of 72 bhp. The new X range replaced the ageing J C B 2D, 3, 3C and 3D Mk. III Excavator/Loaders which had been in production for 13 years, (although they had been continually updated over the years). The 3CX machines had started off at an Excavator/Loader from scratch using anything the design team thought would be better than the current Mk. III models without having to use them as a basis. The Mk. II had only been in production for three years when Project 200 was set up, but even so it still took ten years to design and test a replacement. The 3CX Excavator/Loaders were far superior to Mk. III machines and driver cab comforts were improved to appeal to the export markets. The 3CX was introduced during the slump years when the construction industry was almost at a stand still, and many thought it foolhardy to spend a large amount of money re-tooling the factory for production of the Excavator/Loaders with the prospect of only a small market response.

The new 3CX machines were painted yellow with white cabs, red wheels and front buckets.

When introduced the price of the 3CX was £21,000 and this was not increased for two years after production started. Maximum digging depth of the new JCB 3CX was 14' 2" with a maximum reach of 16' 3", and it weighed just over 6 tons.

The diesel engine powered a power shuttle/reverser transmission with hydraulically operated epicyclic reverser unit, this was operated by a foot pedal for disconnecting engine drive and also gear changing. The 3CX had a four speed full synchromesh unit for on the move gear changes.

Compared with the 3C Mk. III speed of the new 3CX was considerably slower, (especially on hills) the low road travel speed was one of the few faults on the machine (although most drivers would complain about the ignition 'Buzzer' and the lock on the swivel seat).

The replacement for the JCB 3CX was introduced in September 1991 and will also be known as a 3CX.

JCB 360 degree TRACKLAYING EXCAVATORS (1964 - 1980).

JCB's major development was the JCB 7, first of a long line of tracked 360 degree Excavators. It was based on the American 'Warner and Swasey Hopto' design of full slew 360 degrees Excavators. The JCB 7 was a fully hydraulic crawler excavator powered by a Ford 590E six cylinder diesel engine of 96 bhp at 2,250 rpm. It was available with a ? Cu. yd. grab, introduced experimentally at shows in late 1964, and in full production from January 1965.

A two speed hydrostatic transmission was fitted (forward speed 1 mph - 2½ mph). The JCB 7 weighed 12 tonnes 14½cwt, and had a maximum digging depth of 16' 9" with a load over height of 12'

The JCB 6 360 degree Excavator was introduced in 1966, it was similar to the JCB 7, powered a six cylinder diesel engine. Soon after it was followed by the JCB 6C and 6D which were powered by a six cylinder Perkins 6.354 diesel engine of 106 bhp.

The JCB 7 was updated and renamed the JCB 7C and the Ford engine was changed to a Perkins six cylinder engine. In 1969 the JCB 5C and 7B 360 degrees Excavators were introduced. The 5C was powered by a Perkins 4.248 cylinder diesel engine of 77 bhp at 2,850 rpm.

It also had two stage hydraulics and automatic

brakes. The 5C had a digging depth of 18' 11" with a maximum reach of 26' 8". It was slightly smaller than the J C B 6 and weighed 11 tons 8cwt. The 7B 360 degree Excavator (which is more powerful version of the J C B 6D) was powered by a Perkins 6,354 six cylinder diesel engine of 106 bhp at 2,250 rpm. It had a maximum digging depth of 20' with a maximum reach of 29' 7".

The JCB 6C and 6D 360 degree Excavators were introduced in 1971. They were both identical, but the 6D was fitted with an extra wide undercarriage which gave more stability in long reach and heavy dig situations. The maximum digging depth of the 6C and 6D was 22ft, with a maximum reach 31ft. They were powered by a Perkins six cylinder diesel engine of 126 bhp at 2,250 rpm. Prices in July 1971 were £10,596 for the 6D, and in August 1972 was £11,126.

In 1973 the new JCB 800 range of Excavators were introduced. They were the 806 and 807 both of which were fitted with the new hush-flow exhaust system, dual hydraulics and cruciform under-carriage. The 806 and 807 had a high proportion of interchangeable parts.

The JCB 808, 360 degree Excavator first appeared in 1974. It was a larger version of the 807 and had a 138 bhp six cylinder turbo charged engine, which had a hush-flow exhaust system. The 808 could dig to a maximum of 23ft and had a minimum reach of 33ft. It was 28ft long and weighed 49,995 lbs.

The JCB 805 was introduced in 1977, it was another 360 degree Excavator and bought the 800 series up to four models, the 805 was the smallest and powered by a Perkins 4 cylinder engine of 84 bhp, the maximum reach was 30ft with a maximum digging depth of 16ft, some of the optional extras for the 805 included long or short dippers, standard or heavy undercarriage, river maintenance rig and assorted re-handling grabs.

The updated version of the JCB 805 appeared in 1978, it was the JCB 805B 360 degree Excavator. The model was more powerful than the old 805 and the styling was different, with a better cab. It had a Perkins 6 cylinder diesel engine of 112 bhp, noise levels in the cab were 84 dB. The JCB powerslide boom was available as an extra. The boom was hydraulically clamped to the boom base and by releasing the clamp lever in the cab, and using the dipper to extend the boom it could be positioned anywhere over a distance of 3ft. This was done without the driver leaving the cab. The 805B could use either JCB or Hymac buckets.

JCB LOADING SHOVELS 1949 - 1980.

Early Loading Shovels.

The JCB Loading Shovel can be traced back to the Major loader (which is detailed at the front of the book). The Si-draulic and Loadover followed in 1953. The Loadall 75 made an impressive impact on the front loader market. From then on Joe built loading shovels which were based on his JCB 180 degree Excavator/Loader machines. They were the Mk.I J C B 4L, 3S 3CS and 2BS. They were all identical to the Excavator/Loaders, but instead of a rear mounted excavator they had a heavy counter weight. Also available to replace the front bucket were attachments such as a fork lift, crane, concrete skip and road brush.

JCB MKII LOADING SHOVELS.

The JCB Mk. II loading shovels were introduced along side the Excavator/Loaders in 1968. They were the JCB 2BS, 3S, 3CS and 3DS all had the same specifications as the Excavator/Loaders.

A torque convertor transmission was available on all models at £285. After about eighteen months the JCB 3S was dropped from the line up, and the 3DS was renamed the JCB 700, and the prices were as follows:

JCB LOADING SHOVELS - PRICES.

JCB 2BS	£2,058
JCB 3S	£2,190
JCB 3CS	£2,465
JCB 3DS	£2,557

After JCB had taken over Chaseside in 1968 they set about designing a new JCB Loading/Shovel to replace the JCB Chaseside merger range.. The new model was J C B 1250 it was very similar to the JCB/Chaseside 1750, and it cost £8,000. The 1250 was introduced in 1970 as a stopgap until the JCB 400 range of loading shovels were ready for production, the new 400 loading shovels were based around the Chaseside machines, and in 1972 the first of the JCB 400 series Loader/Shovels appeared, these were the result of 5 years research and development, the JCB 413 and 418 were of 4 wheel drive, articulated design and had the engine mounted on the rear half, with the cab and loader on the front, in 1974 they were joined by the JCB 423 and 428, (which were larger versions of the 413 and 418) all 400 machines were fitted with the J C B 'hush flow' exhaust system.

JCB 100 SERIES CRAWLER/LOADERS.

JCB first entered into the Crawler/Loader market in 1971 with the JCB 110 it was a rear engined hydrostatic Crawler/Loader. The engine was mounted rear-ward to give better weight distribution when the bucket was loaded. The JCB 110 was the first Crawler/Loader in production with a fully hydrostatic transmission and won JCB Research a design award in 1973. It was powered by a Perkins four cylinder diesel engine of 73 bhp at 2,250 rpm. It was fitted with a hush-flow exhaust similar to the 360 degrees Excavator. The JCB 110 weighed 21,782 lbs and was 19ft long. It was soon followed by the 112 in July 1975 and the 114 Crawler/Loaders, in January 1976, these were powered by a Perkins six cylinder diesel engine of 101 bhp and 126 bhp respectively. At the same time the 110 was updated to 73 bhp and renamed the 110B.

Sales of the loaders were not very good at first, possibly because the machines were 5 - 10 years ahead of the competition, (the 1985 model 'Lebherr 631' and Caterpillar 963 are similar in design and appearance).

The 100 series Crawler/Loader remained in production until 1978/79 when production was ceased.

J C B TELESCOPIC HANDLERS.

In 1973 J C B Research started testing a prototype Rough Terrain Telescopic Handler, the design team had noted that sales of the bucket mounted fork lifts for Excavator/Loaders were becoming more and more popular, when the move away from bucket fork lifts to specialist fork lift handling machines started. (These were strengthened versions of industrial fork lifts, and as such were not ideally suited to the job at hand). JCB decided to produce a machine purpose designed to perform all the material handling tasks on a construction site, after much market research and a long development programme involving the production and testing of several prototype machines, the JCB 520 was a two wheel drive machine which could lift 2.25 tonnes up to a height of 6.34M (20' 9"). The load was retained on a telescopic boom, and had a good forward reach.

The telescopic boom idea was well received, but the JCB 520 itself was plagued by transmission problems and lack of traction but despite this a total of 422 machines were built, and sold over a two and a half year period, (this represents approximately 14 JCB 520's per month).

The 520 was originally aimed at the construction market, but sales were slow at first, and of the many new markets the JCB tried for their rough terrain handlers, the agricultural market was the best. The JCB 520 was powered by a Leyland 72 bhp engine of the same type used in the 180 degrees Excavator/Loaders.

The JCB 520 was replaced in April 1980 by the 525, which could lift 2.5 tonnes up to 6.4M (21ft). The troublesome transmission of the 520 was replaced by the JCB 'Powertrain' as used on the 180 degree Excavator/Loader. Prices for April 1980 were £14,000 for the 520 and £16,898 for the 520-4.

In June 1980 a 4 wheel drive version of the 525 was launched. The 4 wheel drive eliminated all the previously encountered traction problems of the 520, a workload indicator was added, this worked via. a relay system on the back axle, when the forks are extended the relay lights up 10 lights, one by one, in the cab. The radiator and batteries were repositioned.

MISCELLANEOUS

J C B HERITAGE.

Over the past few years Anthony Bamfords keen interest in preserving his companies past has resulted in a nice collection of some old JCB products from the 50's and 60's being preserved in various parts of the factory. In the foyer of the factory the first JCB trailer, built in 1945, together with a Ferguson tractor and JCB side mower and Joe Bamfords first welding set are on display fully restored. A large collection of Bamfords Ltd. of Uttoxeter machinery is also on display, and includes many engines, ploughs and horse drawn hay mowers etc., which have again been restored by JCB apprentices. A quite unique collection of early machinery and prototype Excavator/Loaders has also been allocated space in a quiet corner of the factory, machines such as JCB Si-draulic loader and fork fitted to a Ferguson TEA20 tractor, a fully restored MKI Excavator and Major loader fitted to a 1955 Fordson Major diesel tractor (rescued from a Yorkshire farm in 1986). A 1953 JCB Hydraulic trencher, fore runner of the MKI Excavator which is awaiting restoration, together with a prototype for the JCB 3CX and a prototype J C B 202 Excavator/Loader, (a smaller version of the 3CX which never went into production). Although not on display at the present, it is likely when room becomes available they will.

BAMFORDS LTD. OF UTTOXETER.

When Joe Bamford left the family owned firm of Bamfords Ltd. in 1945 to set up in business on his own, the old family company never achieved the success that J C B did during the 60's and 70's. In November 1967 Joe Bamford attempted to gain control and take over Bamfords, (who at the time had valuable factory space in the centre of Uttoxeter), unfortunately his take over bid was rejected by the board and major share holders.

I say unfortunately because Bamfords has since closed down and one wonders what JCB would have made out of the old Bamfords company (JCB special products division now owns the former Bamfords west works and are at present building the JCB 2CX and fork lift trucks there.)

J C B PROTOTYPE AND ONE OFFS.

During the last 35 years JCB has built, tested and discarded many new designs and ideas, (some like the '1959 J C B DEXCAVATOR' never even got off the drawing board), but a few were made, machines like the JCB 6D Hydra Dumper prototype with 4 wheel drive and articulated steering, built in 1959, but never put into production.

JCB CRANES.

JCB looked at the possibility of building cranes during early 1960, using parts of the 360 degrees Excavator. They built a lorry type chassis on which to mount cranes, but the high cost and low market requirement prompted Joe to drop the idea. (Much to the delight of another crane manufacturer who had seen the prototype and since stated that the JCB crane would have put them out of business).

JCB WHEELED 360 deg. EXCAVATOR.

During the 1960's JCB research tinkered with the 360 deg. Excavator fitting lattice crane boom and also wheeled versions. Joe had foreseen the demand for a wheeled 360 deg. Excavator, which could be driven from site to site by road, but again production was not thought viable, (although Hymac, Hanomag and Poclain now build similar versions).

JCB MINI EXCAVATOR.

The JCB 1 was the result of putting a small scaled down version of the 'Hydradigger' onto a dumper type chassis during 1959, by 1962 the design was ready and was marketed on the JCB 1.

JCB Serial Numbers and model identification.

Such information as is available is reproduced here, along with details of Fordson Major Serial Numbers.   Over the years records of serial numbers were destroyed. It was also commonplace for manufacturers until the mid 1960s to 'exaggerate' or jump blocks of serial numbers to mislead the competition of how many actual machines were built and sold.

180 degree.

	JCB2/2B	JCB2B/D	JCB3	JCB3C	JCB3D	JCB4	JCB4C/D
June 1962			35002			8115	8153C
Jan. 1963			35875	36092			8434C
June 1963			36577	36594			8756C
Jan. 1964	20003A	20071B	37560	37611			9039C
June 1964	20081A		38587	38835			15519C
Jan. 1965	20237A	20284B	39510	39968			15745C
June 1965	20448A	20409B	40380	40944			15952C
Jan. 1966	20589A	20568B	40627A	41949C			16112C
June 1966	20736A	20774B	42554C				16179C
Jan. 1967	20995A	20997B	43597A	43564C	43710D		16264C
June 1967	21152A	20996B			44446D		16303C
Jan. 1968	21481	21420B	46011A	45937	45904D		47334D
Mk.II			46167	45982	46527		
June 1968		47460B		47485C	47423D		47440D
Jan. 1969		49525B	49520A	49546C	49653D		50760D
Jan. 1970	53274B	56471D	53294A	53261C	53265D		53304D
Jan. 1971	55629B	56673D	56676A	56678C	56651D		56715D
Jan. 1972		589646	59640	59642	59651		
Jan. 1973		65910	65896	65892	65895		
Jan. 1974		103498	103522	103463	103458		
Jan. 1975		108323	108293	108278	108286		
Jan. 1976		111685	111863	111650	111643		
Jan. 1977 (MkIII)		115901	115340	115895	115823		
Jan. 1978		118028	118012	127484	127542		
Jan. 1979		118746	118755	132133	132154		
Jan. 1980		119337	119313	137505	137494		

360 degree.

	JCB5C	JCB6/6C	JCB6C/D	JCB7/7C	JCB7B/C	JCB8C	JCB8D
Jan. 1965				70008A			
June 1965							
Jan. 1966		60003A	60024C	70039A	70201C		
June 1966		60017A		70187A			
Jan. 1967		60021A	60097C	70234	70293C		
June 1967			60203C	70282C			
Jan. 1968		60433A)	60570D	70345			
Mk.II		60562C)			70346C		
June 1968			70397D		70418C		
Jan. 1969	70963C	70659C	70647D	76071C	71008B		
Jan. 1970	71373C	71554C	71380D	71379C	71374B		
Jan. 1971	71684C	71998C	71868D	71904C	71836B	72031C	72327D
Jan. 1972	72373		72374		72345		72360D
Jan. 1973	72983		72879		(72829)(2)		72926
Mar. 1973	73009 (last)		73006 (last)				73087
							(Nov. 1972)

	806	807	808	805
Mar. 1973	75002			
Jan. 1974	75465			
Jan. 1975	76041	76533	76064	
Jan. 1976	76407	76592	76427	76400
Jan. 1977	76759	76757	76786	76799
Jan. 1978	77316	77292	77324	77346
Jan. 1979	77736	77737	77733	77739
Jan. 1980	77994	77991	77990	

Right: Joe Bamford CBE., seen here in 1967 with the welding set he bought in 1945. A new JCB 6C 360 degree excavator and his first trailer are seen in the background.

Left: The JCB logo. This first appeared in 1953 on the JCB Loadover. The lettering gained the style shown when applied to the JCB mid mounted mower, and this was turned round to become the familiar JCB logo, JCB being the initials of Joseph Cyril Bamford.

Below: The first JCB 'Factory. The 'lock-up' garage rented for 30/- (£1.50) a week in 1945. It was demolished by a JCB 6C in 1970.

The original J. C. Bamford trailer sold in 1945 for £45. It is seen here fully restored in the 1970s; it cost Joe considerably more than £45 to buy it back! It is now on display at the factory.

Right: JCB 'Shooting Brake' conversion seen here towing two and four wheel trailers in 1948.

Below: JCB hydraulic tipping trailer (1948). An abundance of ex military components such as wheels and tyres made for economic construction.

Above: The present day factory site. Joe and Son Anthony can be seen by the gate in this 1950/51 shot.

Below: The factory site photographed during the first major rebuilding in 1951/2. Keen eyes will see Nuffield and Fordson Major tractors 'on site'.

Above: J.C. Bamfords 'Major' loader and bucket fitted to a Nuffield Universal M4 tractor. The loader was available with fittings to suit Nuffield or Fordson Major tractors and was built from 1949-57. This photo dates from 1950.

Below: J.C. Bamford 'Master' loader and fork seen fitted to a Fordson Major. it too was available with fittings to suit Fordson or Nuffield tractors. Note the relocation of the tractor headlamps.

Above: J.C. Bamford Half-Tracks fitted to a Nuffield Universal tractor. These units were also available for fitting to Fordson, Ferguson, Ford/Ferguson, Allis Chalmers and David Brown. The conversions proved popular in Scandinavian countries. The photo dates from 1955.

Below: The J.C. Bamford mid-mounted mower built in 1952. Brackets were available to fit it to most popular makes of tractor. The cutter bar was driven via a shaft underneath the tractor transmission and a vee belt from the a pulley attached to the tractor PTO. The design and production rights were sold to Bamfords Ltd. who built many hundreds during the 1950s.

Above: JCB 'Si-Draulic' loader seen fitted to a Ferguson TE-20 tractor. The photograph shows the French version built under licence by 'Hydro-Fourche' in 1953

Below: The first JCB 'Loadover' made up from 'Major Loader' parts. This 1951 shot shows the loader assembled on a Fordson Major E27N tractor, with the bucket in the lifting position after filling from the rear. The unit could either load from the front and discharge at the rear, or vice-versa.

Right: JCB 'Loadover' with the bucket in the discharge position, after lifting over the tractor.

Below: 1953 JCB 'Loadover', production version, built onto a New Fordson Major tractor. The new JCB logo is now applied and obviously hand painted.

Above: A 1953 JCB 'Loadover' shown working in a Gas works yard. The machine was supplied by Knutsford Motors, and was one of six built. The colour of the machine was yellow with red wheels and shovel. The extension on the bucket was to facilitate the charging of the gasworks retorts.

Below: This 1953 view shows a 'Loadover' at Widnes, Cheshire, handling coal; note the merchant's scales in the background.

The **JCB** HYDRAULIC EXCAVATOR

J. C. BAMFORD

Tel.
ROCESTER
283.

ROCESTER - UTTOXETER - STAFFS.
OUR ONLY ADDRESS—NO CONNECTION WITH ANY OTHER FIRM

Grams
LAKESIDE
ROCESTER

This is a very rare copy of the 'first' JCB hydraulic excavator sales brochure, dating from early 1953, showing the unit fitted to a New Fordson Major. It could also be supplied with fittings for the Old Fordson Major, Nuffield, and Ferguson tractors.

Above: The first JCB hydraulic excavator at work in 1953. A choice of 10", 18" or 24" buckets was available.

Below: The same machine shown in the transport position. This excavator was the first JCB hydraulic excavator to be built and marketed but was superseded within a year by the JCB Mk1. Excavator, which is usually quoted as being the first JCB excavator).

The JCB HYDRAULIC EXCAVATOR

J. C. BAMFORD

Tel. ROCESTER 371

ROCESTER · UTTOXETER · STAFFS. · ENG.

Grams LAKESIDE ROCESTER

Copy of the early 1954 Mk1 Excavator Sales brochure.

Above: The production model JCB Mk1 excavator as sold to William Press Ltd., fitted to a New Fordson Major Tractor, trenching across the river Trent in 1955.

Left: Price list/quotation for the JCB Mk1 excavator and Major Loader 1956. The machine was in production from 1954 to March 1957.

Telephone: Rocester 283

Telegrams: Lakeside, Rocester

Your Ref.

Our Ref.

J. C. BAMFORD
Agricultural Engineers
LAKESIDE WORKS
Rocester, Uttoxeter
STAFFS.

COMPLETE PRICE QUOTATION FOR THE

J.C.B. HYDRAULIC EXCAVATOR AND TRACTOR UNIT

One Fordson Major Diesel Tractor, electric
starting and lighting, hand brake, heavy
duty 750 X 16 front wheel equipment, lower
side arm links, counter weight, cab,
J.C.B. Hydraulic Excavator, 18" and 24"
Buckets inclusive of fitting, testing and supply of
J.C.B. Hydraulic Fluid.
Ready for work...................... £1,546. 17. 6d. Ex-Works.

Quotation with above specification and
including J.C.B. Industrial Major Loader
and Bulldozer attachment. Works fitted
and tested........................£1,833. 7. 6. Ex-Works.

TERMS - PRO FORMA

Above: The JCB Mk1 excavator and Major loader, fitted to a New Fordson Major with optional cab, loading a dumper on site. Note the optional 14x30" rear and 7.50x16" front tyres supplied on the tractor ex factory. This indicates a dealer fitted machine, and not a JCB assembled unit. The colour scheme here would be blue tractor with orange wheels and red excavator and loader. Factory assembled machines would have a yellow tractor unit and red wheels, loader and excavator.

Below: This JCB Mk1 excavator has had the front counterbalance weight replaced by a Holman compressor driven from the belt pulley and obviously destined for road breaking/trenching work.

Above: JCB 'Hydra-Digga'/Major loader fitted with JCB cab and shovel mounted dozer blade. The price for the machine shown in 1957 was £1490.0.0.

Below: The JCB 'Hydra-Digga'. The machine had originally been named the 'Diggall' but was changed to 'Hydra-Digga' just before introduction in March/April 1957.

Above: JCB Hydra-Digga/International B450 tractor with front counterbalance weight, produced by JCB and sold exclusively by the International Harvester Company of Great Britain.

Below: A JCB 'Loadall' fitted to a Ferguson TE-20 shown here site stripping top soil in 1956.

Above: An early model JCB 'Hydra-Digga/Loadall' with JCB cab shown at the factory in April 1957. Note the use of the Fordson tinwork, and the earlier design of loader arms and mounting frame.

Below: A Later model JCB 'Hydra-Digga/Loadall' 65. Note the heavy duty loader arms and chassis with JCB tinwork on the bonnet. The rear wheels were fitted the wrong way round to enhance grip in wet site conditions.

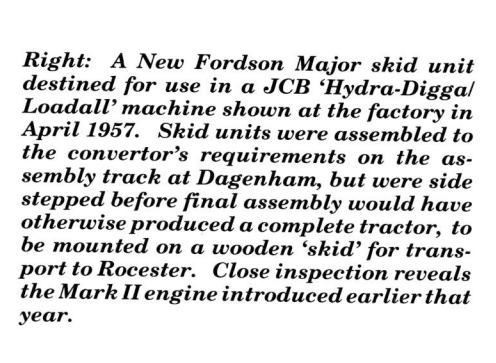

Right: A New Fordson Major skid unit destined for use in a JCB 'Hydra-Digga/ Loadall' machine shown at the factory in April 1957. Skid units were assembled to the convertor's requirements on the assembly track at Dagenham, but were side stepped before final assembly would have otherwise produced a complete tractor, to be mounted on a wooden 'skid' for transport to Rocester. Close inspection reveals the Mark II engine introduced earlier that year.

Left: The rear view of a JCB 'Hydra-Digga/ Loadall' showing reach and load over height. Note the extended 'A' frame struts.

JCB 'Hydra-Digga/ Counter weight, packed up ready for dispatch overseas in 1958/9.

Above: JCB 'Hydra-Digga/ Counter weight', showing transport height of machine. The 'A' frame extension would be folded away for movement on the road. The unit is in this case based on a 'Nuffield' Universal tractor.

Left: Another shot of the JCB 'Hydra-Digga/ Loadall' shown ditch cleaning in 1957.

Above: The JCB 4 replaced the 'Hydra-Digga/Loadall' in 1960. An early model is seen here, on demolition work. Note the round mudguards, and single bucket tipping ram; also visible is a dipper extension.

Below: Another early model JCB4 Excavator/Loader with dipper extension, loading a Bedford tipper.

Left: A Fordson skid unit being turned into a JCB4. The 'A' frame lift ram is fitted. A production line shot from 1960/62.

Right: A later, 1962 version of JCB4 with 'Ripper Tooth' attached. Note the square mudguards and twin bucket tipping rams and the 'squared off' cab.

Left: A 1960 JCB4 showing the inside of cab, back actor attachment and controls.

Above: The later model JCB4 with later type cab. An early JCB4L can be seen in the background which gives a comparison of the early and late models.

Left: A later JCB4 seen at work trenching on the M6 motorway.

Below: The first model JCB3 seen in November 1961.

Above: The JCB3 was the forerunner of todays 180 degree excavator/loaders and had the new JCB 'Hydraslide' which allowed the back actor to be moved along a set of rails to be offset left, right or centre. The JCB was a smaller version of the JCB4. Note the round mudguards with the JCB logo on the front end and the round shovel. The unit shown was a pre-production one for use at the 1961 Crystal Palace plant show on the JCB stand and for show and trial use.

Below: A later model JCB3 Excavator/Loader shown site stripping topsoil. Note the square mudguards and square shaped bucket. These features were introduced in mid-1962.

Above: A wide angle view of a building site in 1962 showing a later model JCB3 excavating footings. Note the old Whitlock 'Dinkum Digger' in the background.

Below: A later JCB3 digging a soak-away in 1962.

Right: A JCB shown trenching in 'competition' with a Massey-Ferguson 2203 Digger/Loader in 1962/3.

Right: The JCB factory in February 1962 with JCB3 and JCB4 production lines in full swing.

Above: The JCB1 rear Excavator/Dozer Blade machine was introduced in 1962. Only 700 had been built by 1966 when production ceased. Most were exported.

Right: The JCB1 excavating a drain in 1963. Power was supplied by a 20HP Petter diesel engine; the cab shown was an optional extra.

Above: Another typical sixties scene with a JCB1 trenching for drainage pipes.

Below: Many of the JCB1s sold in the UK found their way into grave digging service. The JCB1 proved so successful at this work that it was known as the JCB 'Little Grave digger'.

Above: An early model JCB3C shown outside the factory in January 1963. On the first year's production of the JCB3C the only door was at the rear of the cab, but from March 1964 a side door and step was built into the left side of the cab, to allow easier access.

Below: The JCB3C showing Hydraslide rails and kingpost. The level link bar is fitted to this example and was available as an optional extra. the level link was primarily designed for use on soft ground to counteract the tendency for the stabiliser feet to sink in. This accessory cost £49 in 1964.

Above: The JCB was powered by a Fordson Super Major skid unit of 52 BHP and was soon to become JCB's best selling model. Note the rear entry cab and serial number plate on the Rh. mudguard. The serial plate was moved to the inside of the cab on post 1964 models.

Below: The JCB3C at work; note the dozer blade attachment and optional sign board over the mudguard.

Right: The JCB3C seen site stripping with a bucket mounted dozer blade.

Left: The JCB3C with dipper extension and ditching bucket seen ditch cleaning.

Right: The JCB3C shown here in competition with an Essex built Whitlock 60A/W66 Digger/Loader.

Above: The JCB2 Excavator/Loader was powered by a Fordson Super Major skid unit when originally introduced in 1963. It was a smaller version of the 3 and 3C and incorporated the same style of side shift back actor. The later version, the 2B, was identical except for the back actor which was detachable via quick release hydraulic couplings and mounting pins. The skid unit could be factory fitted with tractor linkage, PTO, and drawbar, for use by farmers, contractors, or council/local authorities.

Below: The JCB2 showing the lighter style back actor and hydraslide.

Above: The JCB2 seen loading a lorry on a motorway construction site.

Below: An interesting shot of the JCB range of Excavator/Loaders showing JCB1, 2, 3, 3C and 4C with depth of dig clearly visible. The JCB2 shown is a pre-production model.

Above: The full range of JCB1D dumpers, first introduced in 1964. These two wheel drive manual tip dumpers were available in 30cwt, 22cwt, and 15cwt versions.

Below: The JCB1D 22cwt dumper.

Right: The JCB1D 22cwt. dumper showing engine and driving position.

Left: The engine and skip mounting of the JCB1D 22cwt. dumper. The engine was a 'Bamfords' AC1 air-cooled unit. The rocker covers with JCB emblem were specially cast for these engines.

Right: The JCB 15cwt. dumper.

Above: The Fordson Super Major tractor skid unit shown at the start of the JCB Excavator/Loader production line.

Below: The BMC Nuffield 3/45 skid unit with 3 cylinder engine, shown fitted with hydraulic lift and PTO for use on the JCB2B MkII Excavator/Loader. The larger version using the 4/65 skid unit with 4 cylinder engine was used on the JCB3, 3C and 4 Excavator/Loaders without linkage and PTO.

Above: The JCB3D shown next to a BMC Nuffield 4/65 skid unit which was the power source of the JCB3D.

Below: The JCB3D seen outside the JCB factory was first introduced in 1967; this version shown dates from 1968.

Above: The JCB 180 degree Excavator/Loader production line in 1967, showing an early model JCB3D and a JCB3C behind.

Below: A rare JCB4D Excavator/Loader showing centre mounted back actor. Only a few were built between 1967 and 1971, most going for export.

Above: The last version of the JCB3C Mk1, using the Nuffield 4/65 skid unit. This 1968 model features the optional Drott 4-in-one Clam-shovel.

Below: The 1964 JCB3; a restyled version of the original JCB3 which incorporated a number of improvements to the back actor, slewing, and front shovel. The styling changes are also clearly visible.

Above: The JCB3C MkII, introduced in 1968. This is an early model, note the sloping cab, and tool compartment above door step.

Below: The JCB3C MkII loading a Thames Trader tipper. This 1968 rear view shows the new style back actor.

Right: An early production model JCB3C MkII.

Left: A 1968 pre-production JCB3C MkII showing the Mk1 back actor, and round head and work lights.

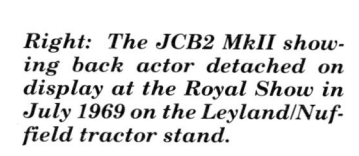

Right: The JCB2 MkII showing back actor detached on display at the Royal Show in July 1969 on the Leyland/Nuffield tractor stand.

Above: The JCB range of Excavator/Loaders and loading shovels in 1970, showing JCB3C II, JCB Chaseside 704, and JCB2DS.

Right: The JCB 180 degree Excavator/Loader line seen in 1969/70.

Below: The 1973 JCB3C II model with white cab.

Above: A JCB3C II 'Dancing Digger' 1973.

Below: The JCB3C with competition in 1971. The machines are, from left to right: Massey Ferguson 2202, Ford 'Auto Dig', Whitlock, and Case Excavator/Loader S.

Above: The JCB3C MkIII shown fitted with a JCB 'Jaw Bucket' similar to a Drott 4 in 1 clam shovel. With increasing competition from other manufacturers by the seventies the 'Another genuine JCB' slogan is of note.

Below: A JCB3C MkIII 1977-1980, shown here in January 1987 clearing snow from Minster Hospital, Kent. Note shovel mounted fork lift attachment.

JCB - DEXCAVATOR

PRELIMINARY SPECIFICATION OF J.C.B. DEXCAVATORS

Excavator	Digging depth	11' 0".
	Reach from King Post.	15' 3".
	Discharge height.	8' 8"
	Ground clearance.	1' 1".
	King Post 3 position:-	Centre and 2' 10½" either side.
	Slew.	180° from centre position.
		200° from offset position.
	Tear-out force	5,650 lbs.
	Height in travelling position	12' 0"
	Width in travelling position.	7' 0"
	Length in travelling position. (with loader)	23' 6".
Loader	Height raised	12' 3".
	Dump height	8' 6"
	Max height to load over	9' 9"
	Max forward reach	3' 3"
	Forward reach at max height.	2' 3".
	Angle of dump	40°.
	Angle of crowd.	15°.
	Tear-out force	6,000 lbs.
	Lifting capacity	1 ton.

HYDRAULICS

Pump	Vickers Vane type. 20 G.P.M. 1,750 p.s.i.		Rams	All rams 3½" bore x 2" dia shafts.
Pump Drive	J C.B. Oil capacity 25 gallons.		Power Unit	Fordson.
Control Valve	Dexcavator:- 6 section 2 lever control Loader:- 2 section.		Tyres	Rear 13.00 x 24 6 ply. Front 7.50 x 16 8 ply.
Hoses	2 wire.			
Filters	Full flow. TARGET PRICE £1900 COMPLETE		Front Axle	J.C.B. Heavy duty with Power Steering.

Above: A rare drawing of the 1959 JCB 'Dexcavator' with preliminary specification and target price. This machine never got off the drawing board, but you can see early thoughts towards the JCB 'Hydraslide'.

Left: The JCB7 was Joe Bamford's first 360 degree tracklaying excavator. The JCB7 had taken him three years to design and get into production. It was powered by a Ford 590E, 6 cylinder diesel engine of 96HP. This was a pre-production model shown in 1964, the model went into production in January 1965.

Right: The JCB7 1965 model shows changes to the cab roof and boom/dipper. The JCB7 was replaced in May 1966 by the JCB6.

Left: The JCB6C was introduced in July 1966. This model could use a 5/8 Cu. yd. bucket at over 20' digging depth and had a 30' reach. It was powered by a 106HP Perkins diesel engine. A 1971 model is shown; the JCB807 replaced the JCB6C in 1973.

Right: The JCB4L Loading Shovel was introduced in 1960 along with the JCB4 Excavator/Loader.

Left: An early model JCB4L seen on refuse tip work. Note the rear door, counterweight, and dozer blade attachment.

Right: The rear view of the JCB4L showing round mudguards, single bucket, and 'Loadall' type shovel arms.

Left: The JCB4CS which replaced the 4L in 1962. Note the square mudguards, new cab and bonnet, and new 'lower style' loader arms.

Right: A JCB/Chaseside Loading Shovel with 4 in 1 Drott bucket on refuse tip work in Surrey in 1969.

Left: A JCB 418 Loading shovel seen in 1970 next to a Chaseside rope shovel based on the Fordson 'N' model restored by JCB factory apprentices.

Right: The most famous machine from the Chaseside legacy was the rope operated loading shovel based on the Old Fordson Major (E27N). These were built from 1945 to 1952. The version shown is a 'Hi-Lift' model also fitted with dozer blade.

Left: The JCB/Chaseside 704 loading shovel.

Right: The JCB3DS Loading Shovel based on a JCB3D Excavator/Loader, seen in 1971. A bulk shovel is fitted. The model was later renamed the JCB700.

Left: The JCB110 crawler Loading Shovel. A prototype known as the TS4 was built from 1968, using a Perkins 4 cylinder diesel engine and track running gear from a JCB5C. The 110 was later joined by the 112 and 114.

Right: The JCB114 on refuse tip work; the engine was rear mounted to aid stability.

Left: The JCB112 with 1 3/4 Cu.yd. 6 in 1 shovel. Production of all these models ceased in 1978/9.

Right: JCB525 Loadall Telescopic Loader. The first model introduced in 1978 was the 520, a 2 wheel drive telescopic handler originally aimed at the construction market. They were more suited for agricultural work however.

Below: A pre-production JCB 'Loadall' in 1977.

Left: The prototype 6D Dump Truck 1959. It was originally displayed at the Crystal Palace plant show in London, but proved too costly to put into production.

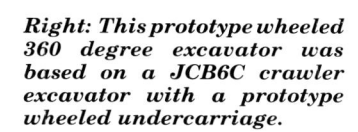

Right: This prototype wheeled 360 degree excavator was based on a JCB6C crawler excavator with a prototype wheeled undercarriage.

Left: Prototype JCB six wheel crane, using a JCB5 crawler excavator mounted on a lorry chassis. When JCB decided against putting the crane into production because of low sales potential and high unit costs one of the UK crane manufacturers told them that they had been very worried about the JCB crane when seen at a demonstration and most relieved when they heard the crane project had been abandoned.

Right: Prototype 'mock-up' of a 'Space Age Digger' designed by 'Brook Stevens' and based on a JCB1C Mk1. The revolutionary round cab was built of wood and had innovatively been designed to give an all-round view for the driver. Joe, however, was not impressed and the cab design was not changed to the futuristic type. The mudguards and new design loader arms were made of steel, but again production was not implemented.

Left: A prototype JCB2 built on a Nuffield Universal 3 tractor on test in 1961/2. Note the strange style of back actor boom which was soon abandoned in favour of the JCB3 type.

Right: JCB3C1 with prototype front axle and oversize wheels and tyres. Note the change of style of the front shovel arms and the snow blade. This machine, one of three built, in 1963/64, is believed to have ended up in Scandinavia.

Left: Early prototype JCB1 Excavator built onto a dumper chassis, seen in 1962.

Right: Another prototype version of the JCB1; note the lack of dozer blade and right hand side low steering/driving position. The back actor is of the later production type.

Below: A JCB2B excavator unit, mounted on an Inland Waterways barge for canal dredging use in 1968. Many similar units were built for the British Waterways Board.

Above: The prototype of the JCB3CX in the late 1970s; note the JCB Mk11 cab and front axle.

Below: A prototype 304, 1975. Many aspects of this machine later found their way into the 3CX introduced in October 1980.

Above Left: Prototype brick hoist on JCB 'Hydradigga' in the late fifties.

Above Right: A prototype crane boom for JCB6 and 6C - late sixties.

Below: Early prototype JCB2 mounted on a Nuffield Universal 3 tractor, compared with the Massey Ferguson 65 tractor and Loader/Backhoe.

Right: A prototype extending dipper on a JCB3C II, 1973.

Left: Prototype JCB4CX Artic-Sitemaster 4x4 development machine seen here in 1988.

Right: Early version of the JCB3CX, October 1980.